7/2012
Aaron
from Grammy

EARTHWORM

Lynn Thomson

123 South Broad Street, Mankato, MN 56001, USA
Printed in Italy
Art Director: Rita Marshall
Book Design: Stephanie Blumenthal
Text Adapted and Edited from the French language by Kitty Benedict
Library of Congress Cataloging-in-Publication Data
Benedict, Kitty.
Earthworm/written by Andrienne Soutter-Perrot; adapted for the American reader
by Kitty Benedict; illustrated by Etienne Delessert.
Summary: An introduction to the physical characteristics, habits,
natural environment, and importance of the earthworm.
ISBN 1-56846-046-5
1. Earthworms—Juvenile literature. [1. Earthworms.]
I. Delessert, Etienne, ill. II. Soutter-Perrot, Andrienne. III. Title.
QL391.A6B37   1992
595.1'46--dc20   92-15024

EARTHWORM

WRITTEN BY

ANDRIENNE SOUTTER-PERROT

ILLUSTRATED BY

ETIENNE DELESSERT

CREATIVE EDITIONS

WHAT IS IT?

An earthworm is a very small animal. When the ground is wet, you may see one crawling along a country lane, or in the grass.

A worm's body is very long and thin. It is formed by many little ringlike segments.

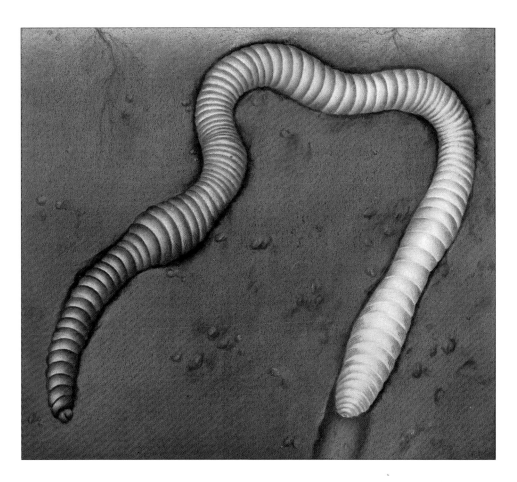

A worm looks the same on both ends, but in fact it has a mouth in front and a small hole in back.

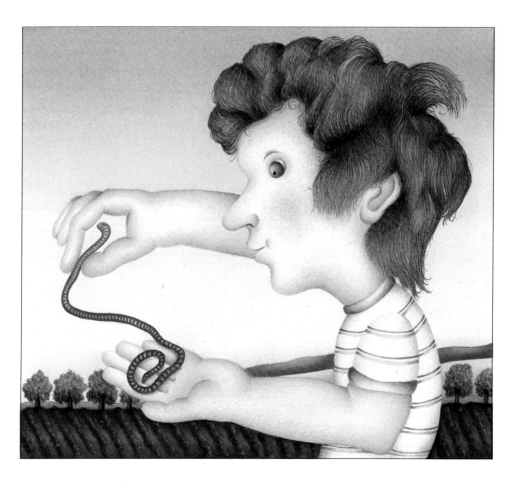

An earthworm is moist and cool to the touch.

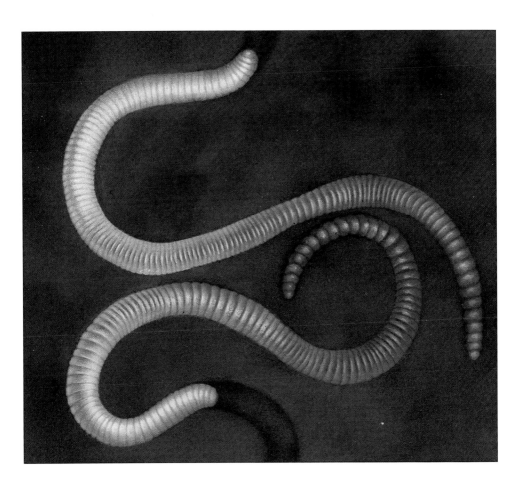

An earthworm is both male and female, but it still needs to mate with another worm before it can lay eggs.

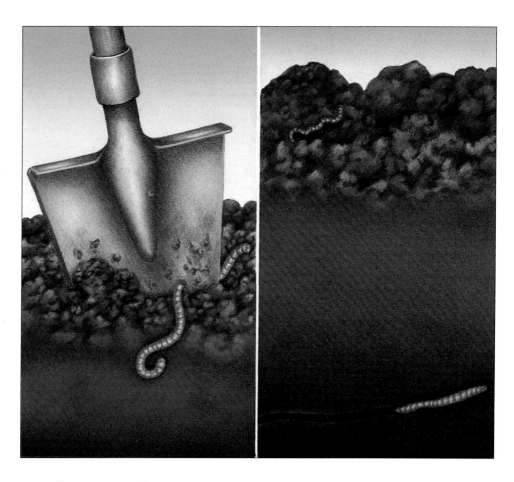

If you cut a worm in two, the halves go on living if they have enough segments left. Each half may grow into a whole worm again. This is called regeneration.

WHERE DO WORMS LIVE?

A worm never lies out in the sun.

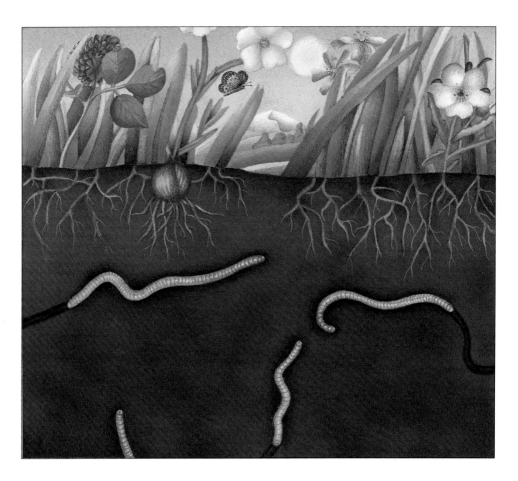

During the day, worms stay deep in the earth, sheltered from the sun's heat and light.

At night, when the air is cool and damp, the worms come up to the surface.

You won't find worms on paved city streets and sidewalks.

You might see worms in your garden or in a park.

In a field, you might find more than five hundred worms in a square yard of soil.

You might find thousands of worms in a compost heap, where grass clippings and garden waste turn into new soil.

Are worms of any use to us?

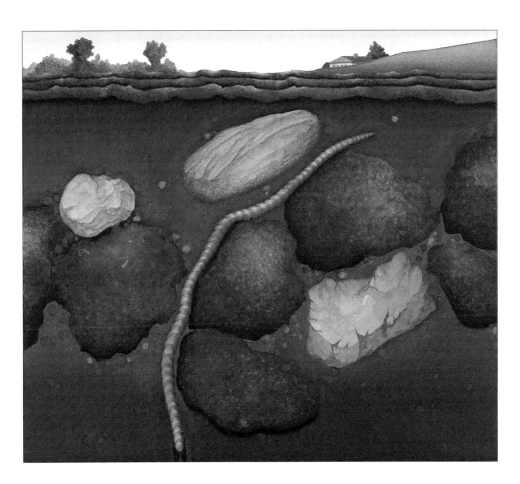

Worms slither between rocks and clods of dirt.

Where the soil is packed down hard, the worms dig out little tunnels, swallowing the soil as they move on.

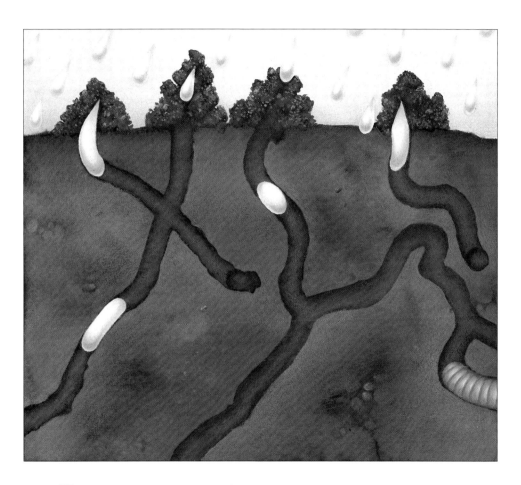

The tunnels let air and water into the ground, helping plants to grow.

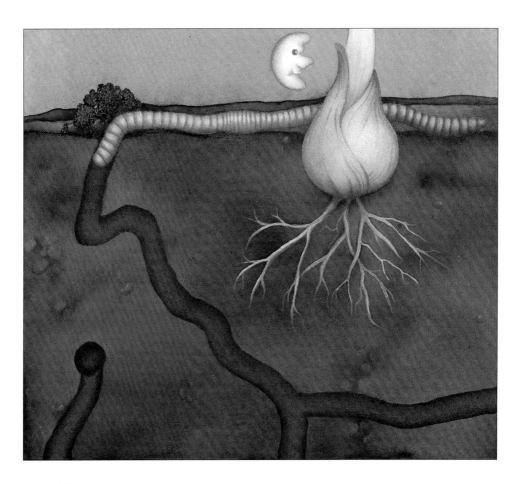

The worms digest the earth they have eaten, leaving behind droppings of finely sifted soil.

These droppings, called wormcasts, make excellent food for the growing plants.

Like a farmer plowing a field, earthworms invisibly till the soil, breaking it up and leaving it finer.

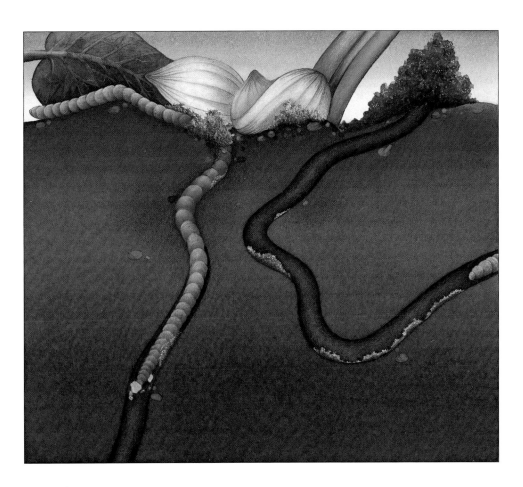

As the worms tunnel underground, they carry bits of grass and leaves with them.

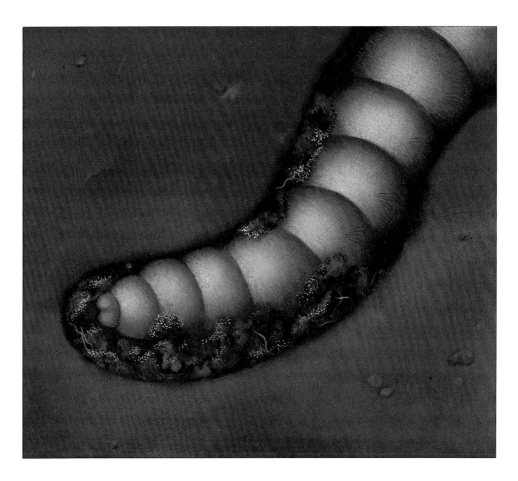

The bits of plants and grass start to rot. Eventually they turn into a dark, rich powder called humus.

All plants need humus, water, and air to grow. The worms help bring the plants what they need.

HOW CAN WE HELP WORMS?

Worms enrich the soil.

They are food for birds, frogs, and many other animals.

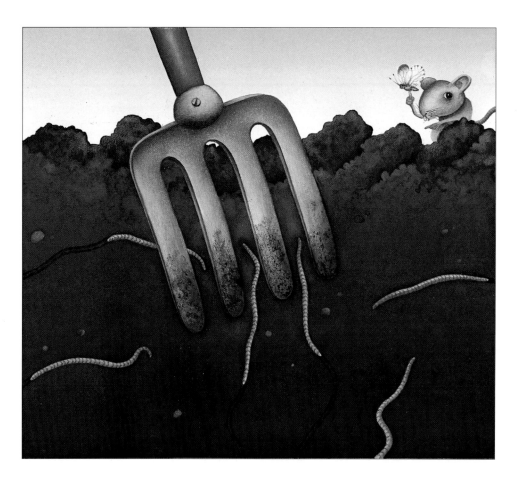

Good gardeners know that worms should not be destroyed. They use gardening tools that will not hurt the worms.

Good gardeners also are careful about using chemicals that might
harm worms or the animals that eat them.

Healthy fields, woods, and gardens always contain plenty of hardworking worms.

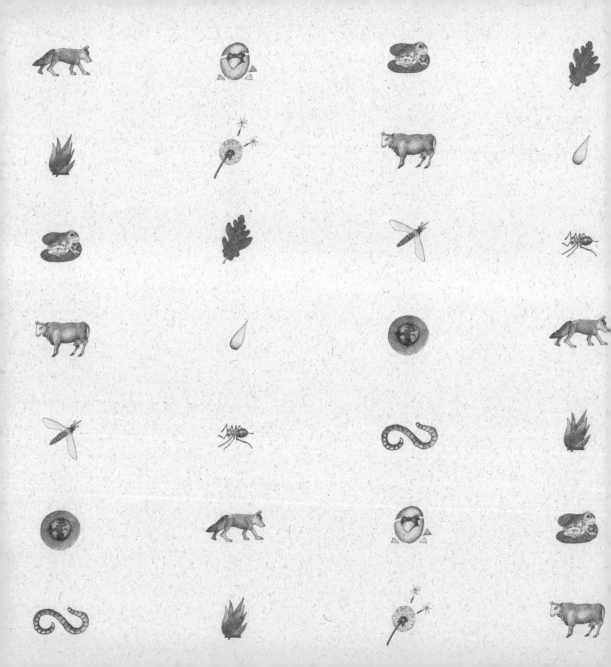